LASERS AND ELECTRO-OPTICS RESEARCH AND TECHNOLOGY

THE DEEP DECOMPOSITION OF WOOD: LIGHT PRODUCTS OF ELECTRON-BEAM FRAGMENTATION

LASERS AND ELECTRO-OPTICS RESEARCH AND TECHNOLOGY

Additional books in this series can be found on Nova's website under the Series tab.

Additional E-books in this series can be found on Nova's website under the E-books tab.

LASERS AND ELECTRO-OPTICS RESEARCH
AND TECHNOLOGY

THE DEEP DECOMPOSITION OF WOOD: LIGHT PRODUCTS OF ELECTRON-BEAM FRAGMENTATION

A. V. PONOMAREV
AND
B. G. ERSHOV

Nova Science Publishers, Inc.
New York

Copyright © 2011 by Nova Science Publishers, Inc.

All rights reserved. No part of this book may be reproduced, stored in a retrieval system or transmitted in any form or by any means: electronic, electrostatic, magnetic, tape, mechanical photocopying, recording or otherwise without the written permission of the Publisher.

For permission to use material from this book please contact us:
Telephone 631-231-7269; Fax 631-231-8175
Web Site: http://www.novapublishers.com

NOTICE TO THE READER

The Publisher has taken reasonable care in the preparation of this book, but makes no expressed or implied warranty of any kind and assumes no responsibility for any errors or omissions. No liability is assumed for incidental or consequential damages in connection with or arising out of information contained in this book. The Publisher shall not be liable for any special, consequential, or exemplary damages resulting, in whole or in part, from the readers' use of, or reliance upon, this material. Any parts of this book based on government reports are so indicated and copyright is claimed for those parts to the extent applicable to compilations of such works.

Independent verification should be sought for any data, advice or recommendations contained in this book. In addition, no responsibility is assumed by the publisher for any injury and/or damage to persons or property arising from any methods, products, instructions, ideas or otherwise contained in this publication.

This publication is designed to provide accurate and authoritative information with regard to the subject matter covered herein. It is sold with the clear understanding that the Publisher is not engaged in rendering legal or any other professional services. If legal or any other expert assistance is required, the services of a competent person should be sought. FROM A DECLARATION OF PARTICIPANTS JOINTLY ADOPTED BY A COMMITTEE OF THE AMERICAN BAR ASSOCIATION AND A COMMITTEE OF PUBLISHERS.

Additional color graphics may be available in the e-book version of this book.

LIBRARY OF CONGRESS CATALOGING-IN-PUBLICATION DATA

Ponomarev, A. V. (Alexander Vladimirovich), 1957-
  The deep decomposition of wood : light products of electron-beam fragmentation / authors, A.V. Ponomarev and B.G. Ershov.
     p. cm.
  Includes bibliographical references and index.
  ISBN 978-1-61728-347-5 (softcover : alk. paper)
  1. Wood--Composition. 2. Wood--Effect of radiation on. 3. Wood--Deterioration. 4. Decomposition (Chemistry) I. Ershov, B. G. (Boris Grigorevich) II. Title.
  TS932.P66 2010
  620.1'22--dc22
                         2010025401

Published by Nova Science Publishers, Inc. † New York

CONTENTS

Annotation	1
Introduction	3
Experimental	7
Distillation of Various Wood	11
Low-Temperature Destruction of Cellulose at Moderate Dose Rates	17
Electron-Beam Distillation of Cellulose	23
Electron-Beam Distillation of Lignin and Binary Mixtures	33
Conclusion	43
Acknowledgment	47
References	49
Index	53

ANNOTATION

Productive direct conversion of lignocelluloses into liquid organic products by electron distillation of crude wood or by dry distillation of preliminarily irradiated wood is considered. Electron-beam conversion is several times more effective than ordinary hydrolytic, pyrogenous or enzymatic processing of wood. Distillation of wood by electron-beam energy yields a number of the useful reagents for heavy organic synthesis and fuel productions. In particular, this path results in high yield of furans - perspective raw materials for production of high-quality alternative fuel which is compatible to conventional engine fuel and the up-to-date types of automobile engines. Electron-beam processing of intermixtures of wood with others synthetic and natural organics (plasts, bitumen, oil, etc.) also can be perspective. An experiments have shown that the electron-beam irradiation of binary wood-bitumen or wood-polymer intermixtures is characterized by sinergistic destruction of components. Such effect can be the innovative base in the alternative fuel production, in recovery of complex polymer-cellulose waste, in processing of native bitumens and heavy petroleum residue, in practical regeneration of synthetic monomers and in synthesis of industrial inhibitors of radical polymerization.

INTRODUCTION

Wood is the important reproducible resource which can play important role in the future strategy of energy safety and a sustainable development. Transformation of wood into liquid and/or gas organic products is considered as a perspective path of raw materials supply for the alternative macroenergetics and for a heavy chemical industry. Reorientation from an oil stock to wood raw materials is caused by a lot of reasons - the accelerated rates of oil and gas consumption; adverse dynamics of the prices; politicization and instability of the hydrocarbons market; non-uniform hydrocarbons deposition; etc. The tendency of the basic economic and ecological coefficients indicates that it is impossible to satisfy future energy and material needs by conserving today's structure of fuel and energy balance.

Fodder grain was used as raw materials for production of firstgeneration biofuel (bioethanol and biosolar oil). Previous experience has shown that one-way orientation to production of biofuel from grain crops is fraught with both a deficit of the foodstuffs and essential dependence on soil fertility, agricultural productivity, weather and climatic parameters. Besides, initially developed processes of cultivation, a harvesting and conversion of grain into biofuel are burdensome enough for environment.

It is supposed that second generation biofuel (grassoline) will be manufactured by ecologically safe methods from wide assortment of inedible raw materials including wood and wood residue. Today's global consumption of oil is about 30 billion barrels per year. Five times more equivalent energy resources can be obtained due to annual global self-increment of inedible cellulose biomass. In-depth research of direct biomass conversion into convenient fuel or into fuel chemicals is the important global problem.

Products of a wood origin are widely used in a daily life. There are many technological processes when wood components are exposed to radiation processing [1-3]. The most known among them are sterilization of medicines and dressing materials; destruction of wood raw materials for the purpose of an intensification of the hydrolytic processing and production of edible, fodder and technical products (blasting powders, lacquers, viscose, various monosaccharoses and agricultural feedstuff from wood); pretreatment of wood for the subsequent manufacturing of various ethers (nitrates, xanthogenates, etc.). The specified aspects stimulate interest to studying of radiolysis of wood and its components, in particular, to detailing of the radiolytic transformations mechanism. At the same time wood components are convenient modelling objects for studying of radiation effects in biological structures.

Electron-beam technologies can play the important role in formation of alternative energy resources from wood. Earlier effects of the accelerated electrons on wood components were investigated widely enough. The ionizing radiation provokes deep chain destruction of wood, changing a molecular-mass distribution and the physicochemical properties (structure, mechanical strength, solubility, reactivity and other.) [3]. The radiation-induced changes facilitate the subsequent mechanical grinding and hydrolysis of wood. These effects are interesting from the point of view of perfection of conventional technologies of wood processing - diminutions of number of technological operations, decrease in power consumption, replacements of ecologically hazardous chemical stages on electrophysical processing without reagents. Previous radiolytic studies have played a key role in determination of the destruction mechanism in lignocelluloses, thereby having accelerated progress of vegetative macromolecules chemistry.

Wood destruction can be intensified by combination of radiation processing with heat processing. Electron accelerators are the basic radiation sources in the modern radiation-chemical technologies. As a rule generated radiation is characterized by high beam current density ~ 100 $\mu A/sm^2$. Processing by such beam results in an effective heating of a material being irradiated (radiation heating). Accordingly, the combination of radiolysis and the fast heating, being the most productive mode of radiation-thermal degradation, can be created by means of only electron accelerator without additional heat sources. The up-to-date accelerators are capable to generate high power electron beams - ≥ 500 kW. It allows to regard the accelerators as the perspective equipment for a large-capacity process industry.

Introduction

The present work views destruction of wood and its main components (cellulose and lignine) at high dose rate of electron beam radiation when radiolysis is accompanied by fast increase of feedstock temperature, initiating distillation (electron-beam distillation) [4-7].

Electron-beam distillation is the updated and advanced modification of a conventional method of dry distillation. The routine dry distillation method (high-temperature airless transformation into volatile products) [2, 8] as well as the enzymatic processing of phytogenous substances are well known long since. Dry distillation of biomass yields both high-quality wood charcoal for metallurgy and a number of valuable chemical reagents - acetone, methanol, acetic acid, phenols, etc. Low molecular weight products of wood distillation can serve as prospective raw materials for synthesis of new polymers. Dominating products of pyrolitic distillation of wood are water (~1/3), carbon oxides (~1/6) and wood charcoal (~2/5) [2, 6]. The total yield of liquid organic products is ≤11 wt %. Condensation of vaporous products results in two immiscible liquids - the transparent light water-organic solution and a dark heavy concentrate of compounds slightly soluble or insoluble in water (tar). High viscosity, density and refraction index of tar are caused by aromatic and furan components. An average molecular weight of tar is ≤ 150. The light water-organic solution contains a small amount of carbonyl compounds and methanol. Slightly yellow color of the solution is induced by furans inclusion. Interest to dry distillation development has weakened owing to both low efficiency of the method and reorientation of metallurgy from charcoal to coke.

EXPERIMENTAL

Anhydrous and airless vegetative samples were used for the experimental electron-beam distillation. Linac U-10-10T (energy, 8 MeV; pulse duration, 6 µs; pulse repetition frequency, 300 Hz; maximal beam current, 800 µA; scan width, 245 mm; and scan frequency, 1 Hz) served as an irradiator (radiation heater).

Experimental electron-beam distillation can be executed by means of the simple equipment. The universal schema of laboratory installation is shown on Figure 1. The electron beam *1* is oriented to the quartz reactor *2* containing a plant material (3-80 grams). Vaporous products are condensed by air condenser *3* (at 17±2°C) and by water cooler *4* (at 15±2°C) outside of irradiation area. Liquid products separate from gases in the collector flask *5*. Gaseous products can be returned again into a reactor *2* or removed from installation by pump *8*. Inner gas composition in installation can be adjusted using a gas bottle *7* (for example, by hydrogen or by light alkanes).

Absorbed dose rate was estimated using aqueous bichromate dosimeter and standard film dosimeter with a phenazine dye-doped copolymer.

The average dose rate was 2 kGy/s. It is necessary to mention that absorbed dose rate was not a stationary value owing to transformation of a feed stock to charcoal. Radiation use factor was 5-25 %. Distillation products (sampled from *3* and *6*) were analyzed chromatographicly (Q-Mass, Perkin Elmer AutoSystem XL; helium as the carrier gas, capillary glass column 60 m long 0.25 mm in inner diameter).

Typical dynamics of an electron-beam heating of wood at dose rate 2 kGy/s is shown on Figure 2. The most intensive condensation of vapours inside coolers is observed after 3 minutes whereas after 6 minutes the vaporization is terminated. Intensive vaporization testifies to the chain mechanism of wood destruction. Observed yield of condensation is ≥ 15 µmol/J. Only wood charcoal remains in a reactor at beam heating above 400°C.

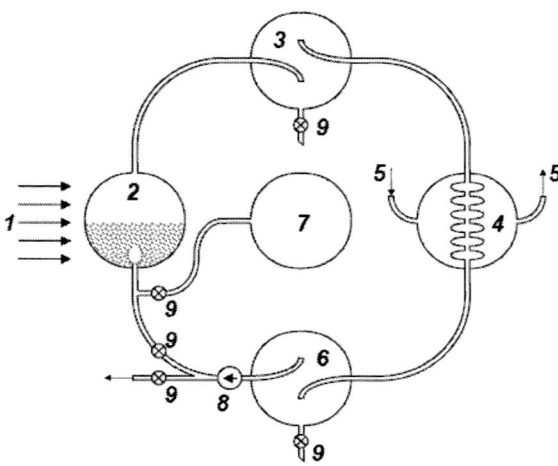

Figure 1. Schematic diagram of electron-beam distillation: (1) electron beam, (2) quartz reactor, (3) air condenser, (4) water cooler, (5) cooling water, (6) collector flask, (7) gas bottle, (8) gas pump and (9) valve.

At atmospheric pressure wood pyrolysis could take place at temperature above 270°C [2]. However the sample under beam irradiation does not reach such temperature a long time (see Figure 2). A fast chain formation and transpiration of radiolytic products preclude an excessive heating in a reactor - the heating curve has sag. Thus the probability of the adverse "pyrogenic" destruction mechanism is minimized. So the self-organization effect (self-preservation mechanism) takes place in electron-beam destruction process. In temperature range 220-260°C the beam irradiation does not provide a fast feedstock heating. Heating rate is retarded testifying to initiation of effective endothermic process. The majority of liquid radiolytic products have boiling points in the range 100-250°C. Their specific evaporation heats are 30-46 kJ/mole. Warmth saved up at initial radiation heating is ~10 times less. Main part of beam energy is used for feedstock fragmentation and evaporation of radiolytic products, not yielding the further heating. On the other hand, apparently, the temperature ≤260°C degrees is sufficient for chain cleavage of instable radiolytic intermediates. Significant rise of temperature is being renewed after wood charcoal formation.

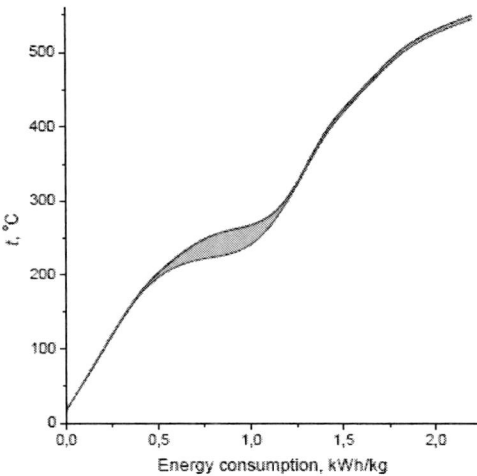

Figure 2. Typical dynamics of electron-beam heating. Data observed at distillation of several wood types are placed between curves (feedstock specific surface $Z=105$ cm^2/g and bulk density 150 g/dm^3).

The accelerated electron loses energy by small portions (~20 eV on the average), forming 2-5 nm spurs - the isolated zones of ionization and excitation [1, 9]. The typical distance between spurs is hundreds nanometers. Recombination of the ionic and radical pairs in a spur results in the energy liberation (equivalent to local fast rise in temperature). Thus the sequence of the high-temperature nano-reactors isolated from each other is promptly being shaped along an electron trajectory. In contrast to pyrolysis the electron-beam radiolysis preferentially produces the excited and super-excited molecules preceding the splitting of skeletal bonds. Usually a singlet excitations decay by forming the molecular products whereas triplet states dissociate into the radicals [1]. Excess energy is being dissipated along cold macromolecular chains, supporting further cleavage of instable intermediates.

DISTILLATION OF VARIOUS WOOD

The influence of the wood type on the phase composition of electron-beam distillation products is shown in Table 1. Softwood yields a somewhat greater amount of charcoal as compared to hardwood. At the same time, the relative amount of charcoal remaining after radiation heating is lower than that upon the conventional pyrogenic treatment. The conventional dry distillation of pine, spruce, alder, and birch produces about 33–36 wt % charcoal and 15–17 wt % incondensable gases; almost a half of the liquid condensate is water [2].

Differences between wood species are manifested already at the step of condensation of the vapors distilled off in the radiation heating mode. Approximately half of vapor produced from the soft wood and less third part from foliar wood are condensed in an air condenser. The fresh condensate contains a small amount of water (≤8 wt %). The water fraction is incremented during storage of the condensate, especially, upon heating which accelerates the water formation. The instability of the condensate assumes necessity to dilute and to process (stabilize) the low-molecular-mass products of distillation already at the vapor withdrawal step [6].

The component composition of the condensable products of electron-beam distillation substantially depends on the type of wood used (Table 2). The condensate obtained upon the destructive distillation of softwoods is enriched in phenolic compounds. Their precursors are lignin and other aromatic wood macromolecules [2]. Aromatic compounds exhibit radioprotector and antioxidant properties [1, 10]; therefore, their increased concentration predetermines a number of significant specific features of degradation of softwood. Softwood condensates contain a smaller amount of low-molecular-mass components composed of 1–3 carbon atoms.

Table 1. Phase composition (wt %) of the products of electron-beam distillation of various wood feedstock (in *i*-butane at atmospheric pressure; feedstock specific surface $Z=40\pm5$ cm^2/g and bulk density 150 g/dm^3).

Wood Feedstock	Charcoal	Gas	Liquid
Alder (Alnus)	25.4	18.0	56.6
Aspen (Populys tremula)	25.4	16.7	57.9
Birch (Betula)	23.7	14.7	61.6
Larch (Larix)	28.9	14.5	56.6
Linden (Tilia)	23.5	15.1	61.4
Mespilus (Amelanchier)	25.0	17.1	57.9
Messmate (Eucalyptus)	28.5	14.0	57.5
Pine (Pinus)	29.6	13.3	57.1
Poplar (Populus)	25.5	16.5	58.0
Spruce (Picea)	29.5	15.9	54.6

This is an indication of less severe fragmentation of wood enriched in aromatic components. Correspondingly, the average molecular mass of pine and spruce condensates is higher than that of hardwood condensates (Table 2). The liquid mixtures of softwood electron-beam distillation products are substantially more stable upon storage and heating. At the same time, a higher yield of charcoal is probably also due to an increased concentration of aromatic components in the softwood feedstock.

Radiation-induced degradation of both softwood and hardwood results in a considerable amount of the furan fraction (Table 2). The main source of furans is cellulose [5]. Among furan derivatives, furaldehydes prevail. Radiation heating of hardwood yields 3-furaldehyde along with 2-furaldehyde (furfural). Furfural becomes a dominating representative of furaldehydes in the case of decomposition of softwood material. Aspen and alder condensates are characterized by a higher proportion of furanmethanols, whereas the condensate from pine contains 5-methyl-2-furaldehyde in a relatively greater amount. However, the variety of furans produced from softwoods is greater as compared to hardwoods. For example, the pine condensate additionally contains furan, 2-methylfuran, 2,5-dihydrofuran, 2,5-dimethylfuran, 2,5-dihydro-3-methylfuran, and 5-methyl-2(3H)-furanone.

Table 2. Component composition (wt %), average density ρ (at 18°C), and average molecular mass M of condensates from wood feedstock (in i-butane at atmospheric pressure; feedstock specific surface $Z=90\pm5$ cm^2/g and bulk density 150 g/dm^3).

Main Components	Wood Feedstock				
	Birch	Aspen	Alder	Spruce	Pine
Methanol + Methylformiat	10.1	10.9	13.9	12.7	9.8
Acetone + propenal	4.3	2.4	3.4	3.4	2.3
2,3-Butanedione	5.9	9.9	7.2	3.8	4.5
2-Oxopropanal	3.0	3.7	4.9	1.2	1.2
Acetic acid	26.1	28.0	23.5	15.1	10.6
1-Hydroxy-2-propanon	6.1	6.2	4.9	9.7	8.6
2-Furaldehyde (furfural)	13.1	11.1	11.2	15.6	16.1
3-Furaldehyde	9.5	2.9	4.8	0.3	0.1
Furanmethanols	1.6	9.0	6.7	3.3	3.1
5-Methyl-2-furaldehyde	0.5	0.3	0.8	0.8	2.0
4-Methylphenol	≤0.1	0.6	0.2	≤0.1	0.3
2-Methoxyphenol	≤0.1	0.1	1.5	3.7	6.1
2-Methoxy-4-methylphenol	≤0.1	0.1	0.2	0.7	4.0
Total furans	24.7±2.1	24.9±2.3	24.8±2.9	24.9±3.9	28.4±5.1
Total aromatics	≤0.3	2.2±0.9	2.6±1.0	13.9±3.0	18.4±6.2
Density ρ, kg/m^3	1141	1122	1119	1138	1125
Mole mass M, kg/mol	73.8±1.1	73.6±0.7	75.8±1.4	79.4±1.6	81.2±1.4

The wide variety of furans and their higher total yield can also be due to a reduction in the severity of cellulose fragmentation in the presence of softwood aromatic components.

The most representative fractions in condensates are low-molecular-mass oxygenated compounds. Oxygen atoms appear in various functional groups, forming acids, aldehydes, ketones, alcohols, ethers, and esters. The relative distribution of compounds bearing different oxygen-containing groups in the total oxygenated components is shown in Table 3. Carbonyl compounds occupy almost 3/4 of the total mass of hardwood condensates, but their fraction in softwood condensates is noticeably lower.

Table 3. Distribution of components in the oxygenate fraction of condensates, wt % (in *i*-butane at atmospheric pressure; feedstock specific surface $Z=40\pm5$ cm^2/g and bulk density 150 g/dm^3).

Functional group (compound)	Wood Feedstock				
	Birch	Aspen	Alder	Spruce	Pine
-OH (alcohols, phenols)	13.8	12.3	8.4	20.2	28.1
-COOH (acids)	41.3	35.8	35.6	22.4	17.8
-C=O (ketones)	11.6	23.1	18.5	19.5	17.0
-C=O (aldehydes)	17.6	16.2	21.3	18.1	22.4
-O- (ethers)	10.5	10.7	12.9	15.1	22.2
-O- (esters)	5.2	1.9	3.3	4.7	0.4

Ambient isobutane too participates in radiolysis, being in particular the predecessor of liquid alkanes. The fraction of liquid alkanes in the condensates is relatively small (≤ 2 wt %) and includes C_8 isomers. On the other hand, there are butanols and some ethers untypical of the products of electron-beam distillation of the wood itself among the oxygen-containing components. It is evident that the formation of these products is associated with radiolytic transformations of the isobutane, the carrier gas. It is interesting that the mode of electron-beam distillation of wood in a light alkane medium provides the yield of the gas alkane fixation on order of magnitude above than radiolysis of dry or watered gas without wood additives [11, 12]. This effect can result from the processes of charge and energy transfer from lower products of chain fragmentation of wood to light alkane molecules. As a consequence, more effective degradation of the gas increases the probability of participation of its intermediates in the formation of liquid alkanes, alcohols, and ethers.

Natural rotting processes have an effect on the product composition of electron-beam distillation of wood [13]. The condensate contains a lesser amount of light products resulting from cellulose fragmentation whereas the fraction of aromatic fragments and furans increases. For example, in the case of electron-beam distillation of rotten pine dust, the relative amount of aromatics and furans in the condensate reaches 65–70 wt %. The total condensate yield decreases to 40 wt % and the yields of charcoal and gas increase to 42 and 18 wt %, respectively, in this case. These changes indicate that aromatic compounds are more stable during natural biochemical wood aging processes.

LOW-TEMPERATURE DESTRUCTION OF CELLULOSE AT MODERATE DOSE RATES

Cellulose forms vegetative cellular walls, it is a main structural component of natural phytogenous matters. Radiation-chemical yield of cellulose destruction at room temperature is $G^0 = 6.0 \pm 1.0$ [3]. Both dose rate and initial polymerization degree and microcrystallinity of cellulose do not practically influence a destruction yield. The dense crystalline structure results in low gas permeability of cellulose fibrils. Therefore the vacuum, ambient inert gases and air also feebly influence cellulose radiolysis. Oxygen promotes only superficial radiolytic oxidizing, forming carbonyl and carboxyl groups on a cellulose surface. Cellulose destruction is a little inhibited by a moisture [14, 15].

Dependence of cellulose polymerization degree P on a dose D at doses <200 kGy is linear, being well featured by the equation (1).

$$\frac{1}{P} + \frac{1}{P^0} = \frac{1.62 G^0 D}{N_A k} \tag{1}$$

where P^0 – initial polymerization degree of cellulose, k - auxiliary coefficient, N_A - Avogadro number. Decomposition rate decreases at higher doses. This effect is caused by formation and accumulation of fragmentation products. These intermediates and final products retard the further destruction of cellulose, being scavengers of electron, radical and excitations. Cellulose destruction at very high doses (> 1 MGy) can be featured by empirical equation (1) in which D should be substituted on D^q where q is equal 0.83 [16, 17] or 0.73 [18].

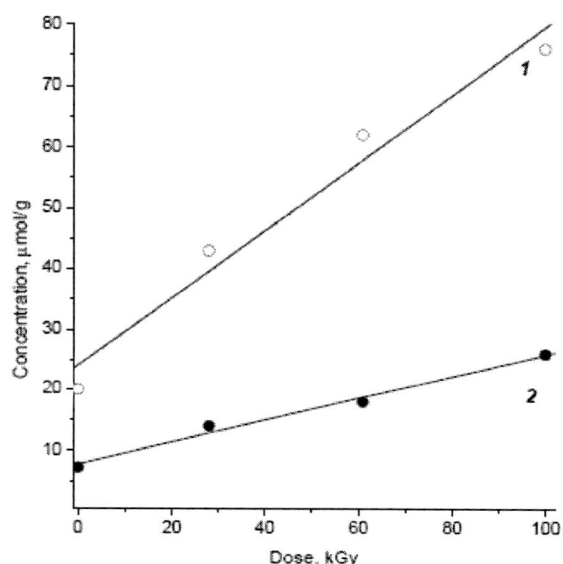

Figure 3. Accumulation of carbonyl (*1*) and carboxyl (*2*) groups in a cotton cellulose [23].

The irradiation at low temperature and at low dose rate results in the deep chemical transformations of cellulose [19-23] Among the water-soluble products the cellobiose, glucose, arabinose, glyoxal, 2-ketogluconic acid and some other acids are found out. Hydrolysis of the water-soluble fraction results in formation of xylose, arabinose, glucuronic and formic acids, malonic dialdehyde, etc. Figure 3 illustrates influence of an irradiation on the content of >C=O and –COOH groups in a cotton cellulose at room temperature.

Preferentially carbonyl compounds are formed via cellulose destruction. Yield of carbonyl groups formation is 6.5 ± 0.5. Yields of carboxyl groups are from 0.9 to 1.8 [18, 24-29]. Radiolytic formation of carbonyl and carboxyl practically does not depend on the initial content of these groups in raw cellulose (see Table 2). About 3 mmoles >C=O and 1 mmole -COOH are formed in 100 g of cellulose at an absorbed dose 50 kGy.

Both carbonyl and carboxyl concentrations at high doses are featured by non-linear dependence $c = BD^q$. The coefficient B is $6.8 \cdot 10^{-7}$ for carbonyl and $1.8 \cdot 10^{-7}$ carboxyl groups [16]. The exponent q for both groups is 0.50-

0.69 [18, 24]. Apparently this effect at high doses is due to an involving of carbonyls and carboxyls into radiolytic transformations.

Hydrogen, dioxide carbon and monoxide carbon are main light-end products of cellulose radiolysis [21, 23, 25, 31-34]. Their yields in vacuum at doses ≤ 300 kGy are 6.0, 3.2 and 0.9 respectively. Methane also has been detected among light-end products.

The conventional view on the nature of the free radicals in irradiated cellulose has been formulated after the publication [15, 35-50]. Predominating radical states in the irradiated cellulose are localized on endgroups of polymeric chain.

Probabilities of C(1)-H and C(4)-H bonds cleavage in glucopyranose at nitrogen boiling point are similar. The irradiated samples of the frozen cellulose have the dark blue colour due to optical absorption of trapped electrons. Colouring is intensified in the moist samples.

Radiolytic destruction of cellulose is initiated by ionization, ion-electron pair neutralization and decay of excited states:

$$RH \rightarrow RH^+ + e^- \qquad (2)$$

$$RH^+ + e^- \rightarrow RH^* \qquad (3)$$

$$RH^* \rightarrow R\cdot + \cdot H \qquad (4)$$

Excited states decay preferentially by cleavage of C(1)-H or C(4)-H bond in glucopyranose ring (radicals RO☐ in the irradiated cellulose are not detected). H· atoms eliminate hydrogen from glucopyranose, forming the same radicals R·

$$RH + \cdot H \rightarrow R\cdot + H_2 \qquad (5)$$

Primary R· radicals, apparently, are instable as a result of the considerable strains induced by discrepancy between electronic configurations of radical centre (sp^2-hybridization) and initial molecular unit (sp^3-hybridization). Already at room temperature R· radicals degrade by glucosidic bond cleavage [3]:

$$\text{(6)}$$

$$\text{(7)}$$

The fragments from (6) and (7) reactions decay to form further stable low-molecular-weight products. The yield of carbonyl groups formation is enough close to a yield of destruction of cellulose and a yield of CO_2 formation; i.e. each cleavage of cellulose polymeric chains results in formation of one molecule CO_2 and one carbonyl group.

Change of temperature from 300 to 77 K results in almost double decrease of cellulose destruction yield, as well as formation yields of carbonyl compounds, CO_2 and CO. In the same conditions the H_2 formation yield is invariable. It indicates that C-H cleavage not depends on temperature, whereas cleavage of glucopyranose units can be retarded by a cooling. Process of radiation destruction of cellulose can be stimulated by a heating [19]. The temperature is more, the degree of polymerization is lower. Temperature effect on cellulose destruction is shown on Figure 4. There are two temperature ranges differing by efficiency of radiation destruction in cellulose. Destruction yields vary slightly at temperatures below 110°C. Higher temperatures result in appreciable increase of the destruction yields.

Activation energy of destruction in high-temperature range (110-190°C) is 23.4 ± 1.6 kJ/mol. Such value is characteristic for diffusion of radicals in the polymer and corresponds to energy of conformational oscillations of glucopyranose unit [51]. The destruction yield coincides with a yield of carbonyl groups formation (see Figure 4). High-temperature destruction of cellulose is also accompanied by increased yield of hydrogen and carbon dioxide.

The radical-chain mechanism of radiation-chemical destruction of cellulose takes place at temperature above 110°C. At the same time ratios of main products yields do not depend on temperature. Hence there is a likeness of processes of a radiolytic fragmentation of cellulose in different temperature ranges - decay of each glucopyranose unit is accompanied by formation of carbonyl compound and CO_2 molecule [3].

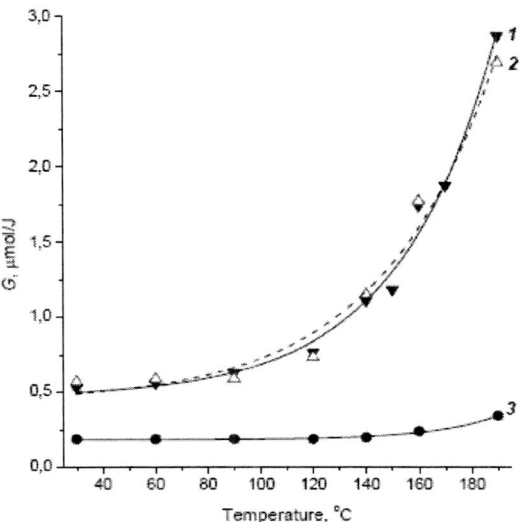

Figure 4. Temperature dependence of destruction yield of cellulose (*1*) and formation of carbonyl (*2*) and carboxyl groups (*3*) [19].

Cellulose destruction depends significantly on operational sequence of an irradiation and heating [52]. Both the irradiation and thermal processing result in decrease in cellulose polymerization degree and in increase of a yield of hydrolytic formation of reducing sugar. Bond breaking in a cellulose macromolecule at sequence "heating→irradiation" corresponds to the equation:

$$S_{(T \to R)} = S_T + S_R \qquad (8)$$

where $S_{(T \to R)}$ - number of cleavages via the consecutive two-stage processing; S_R - number of cleavages via single-stage radiolysis; S_T - number of cleavages via single-stage thermal processing. That is, the sequence "heating→irradiation" yields the additive effect. The sequence "irradiation→heating" yields the superadditive number of cleavages [21, 53, 54]:

$$S_{(R \to T)} = S_R + S_T + \mathrm{B}d \qquad (9)$$

where β - the coefficient corresponding to a rise of efficiency of thermal destruction of irradiated cellulose at dose D. The sequence "irradiation→heating" gives also the superadditive influence on a yield of hydrolytic formation of reducing sugar from the treated cellulose [52].

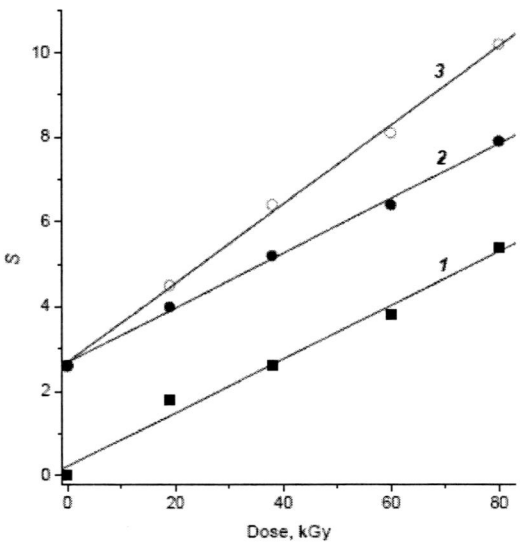

Figure 5. Dose dependences of bond cleavage number (S) in wood cellulose at various combinations of an irradiation and a heating (3 h at 190 °C): 1 – irradiation; 2 - heating→irradiation; 3 - irradiation→heating.

Dose dependences of bond cleavage number in wood cellulose are shown on Figure 5 at various combinations of an irradiation and a heating. The slope of linear dependences *1* and *2* is identical, confirming that pre-award thermal processing does not influence efficiency of the subsequent radiation destruction. At the same time, larger slope of a line *3* displays more effective destruction of cellulose at sequence "irradiation-heating". The coefficient β correlates with slopes variance of dose dependences at specified treatment temperature. At 150, 170 and 190°C β-values are 0.004, 0.02 and 0.038 bond cleavages per 1 kGy [21, 54].

ELECTRON-BEAM DISTILLATION OF CELLULOSE

We studied the efficiency of electron-beam conversion of cellulose into liquid organic compounds depending on initial temperature, dispersion degree (specific surface Z) and feedstock type. Three cellulose types were tested: a cotton cellulose (C1), the unbleached pine sulphate cellulose (C2) and the bleached pine sulphite cellulose (C3).

Telemetry does not reveal any significant distinctions in dynamics of electron-beam distillation of C1, C2 and C3 samples in airless medium at 730-740 mm Hg pressure. Absorption of dose ~250 kGy is accompanied by formation of the heavy fog gradually flowing from a reactor into condensers. Vapours colour changes from initial white to brightly yellow.

The condensate accumulates in a collector flask. Its amount increases during an irradiation. Formation of vapours and build-up of the condensate level are terminated at a dose ~500 kGy. The condensate represents the homogeneous odorous brown liquid; yield of a condensate formation ~60 wt % (see Table 5). The dry black residue (charcoal) remains inside a reactor after distillation. Volume of charcoal is less than initial volume of a feedstock. Charcoal yield is approximately 20 % of initial dry weight of cellulose.

The content of the main components in the condensates formed from C1, C2 and C3 is shown in Table 6. Condensates originated from all three samples have similar compositions and include over 40 organic compounds of molecular weight 32-128. Main liquid products are furans among which 2-furaldegid (furfural), 2-furanmetanol, 5-metil-2-furaldegid and 3-furaldegid predominate. Water content in a condensate is ≤ 8 wt %. Similar results have been revealed earlier [55] by examination of electron-beam distillation of cotton wool and sulphate pulp in a gaseous alkanes flow.

Table 4. Initial concentrations and radiation-chemical yields of the oxidized groups in cellulose (γ-radiolysis) [30].

Cellulose type	Concentration, mmoles/100 g		Yield, mmoles/100 eV	
	-CHO	-COOH	-CHO	-COOH
Wood sulphite cellulose	0.13	0.24	6.0	2.0
Wood sulphite high-viscosity cellulose	0.53	3.16	5.4	1.8
Cotton cellulose	0.10	0.45	5.9	2.5

Table 5. Yield G (in % from dry weight of cellulose), index of refraction n_D^{18}, density ρ^{18}, and viscosity η_0 of condensates (Z=104 cm^2/g).

Feedstock	G, wt %	n_D^{18}	ρ, kg/dm^3	η_0, mPa·s
C1	63	1.4479	1.1639	5.96
C2	58	1.4449	1.1560	5.67
C3	60	1.4455	1.1594	5.79

Table 6. The average content of the main components in condensates (wt %).

Component	Feedstock (Z=104 g/cm^3)		
	C1	C2	C3
Methyl formiate	0.9	0.8	1.3
Acetone	4.9	2.7	3.4
Formic acid	5.0	5.2	3.6
2,3-Butandione	1.6	3.5	1.7
2-Oxopropanal	3.4	0.6	0.8
Acetic acid	7.3	6.0	3.1
1-Hydroxy-2-propanone	7.3	10.7	1.5
1-Methylpropylacetate	1.8	0.9	2.0
Furaldehydes	42.5	44.3	51.0
Furanmethanols	2.0	2.8	4.7
Methylfuraldehydes	9.4	13.3	17.3
Total Furans	72.5	64.4	79.0

Crushing of C2 does almost not influence duration and intensity of electron-beam distillation. The condensate yield does not change - 58±1 wt %. However the light tendency to decrease of a chared residue yield and increase of a volatile products yield (about 2-3 wt %) is observed for more crushed C2 samples. At the same time, composition of a condensate varies - the finer crushing of a feedstock, the more both density, viscosity, average molar mass, refraction index, and optical density of a condensate (see Figures 6 and 7). Following correlations are being detected by means of comparison of the condensates forming from cellulose via a specific surface Z magnification:

> ➤ the fraction of the furan-derivatives representing large fragments of a glucopyranose (such as furaldehydes, furanmethanols, furanones) increases;
> ➤ the fraction of small fragments of a glucopyranose (such as 1-hydroxy-2-propanone, acetic and formic acids) decreases;
> ➤ the fraction of the heavy products forming in the secondary interactions between main primary products (such as furanmethanolacetates) decreases also.

Undoubtedly, products of a radiolytic fragmentation of cellulose form both on a surface and in volume of solid particles. Migration of fragments from volume to a surface (into the mobile vaporous phase) demands a time. That delay increments probability of additional radiolytic destruction of primary fragments or their reactions with other products (formation of "hybrid" compounds). Apparently, such effects result in distinction of composition of the condensates received at distillation of samples with various crushing.

As a rule, cellulose pyrolysis is initiated at ~270°C. It would seem that radiation heating of hotter samples should result in to more prompt distillation. However it does not occur.

Influence of a preliminary heating on electron-beam distillation was studied at temperature ≤250°C to prevent cellulose pyrogenic decomposition before an irradiation [2]. Insignificant temperature effect on a condensate yield was observed. According to telemetry the observed time of the vapors emersion in the preheated samples has decreased a little, however end time of distillation did not depend on initial temperature. It specifies that dynamics of the electron-beam distillation is controlled by radiolysis and radiation heating, instead of feedstock temperature. The condensate yield at initial feedstock temperature 16-250°C was 58.0±1.5 wt %.

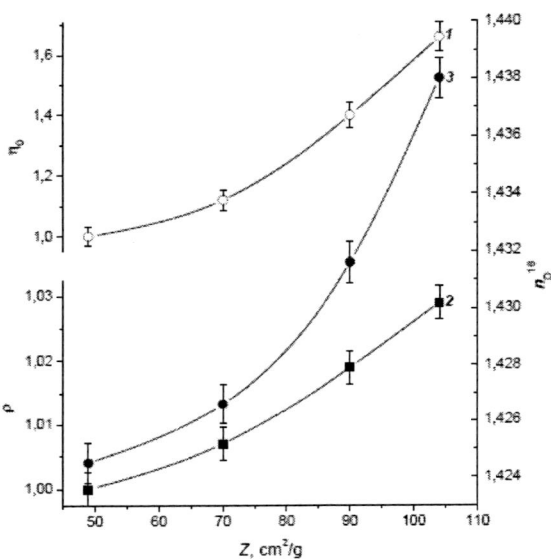

Figure 6. Influence of specific surface Z of samples C2 on the relative values of density ρ (1), viscosity η_0 (2) and refraction index n_D^{16} (3) of condensate.

Figures 7 and 8 testify that the preliminary heating promotes formation of heavier condensate - while raising an initial feedstock temperature, an increase of density, viscosity and index of refraction of a condensate are observed. The molar mass of a condensate continuously grows from ~87 to ~94 g/mol. Optical density at λ_{max}=275 nm (typical of furans) increases. Furans-derivatives fraction in a condensate expands from 60 to 72 wt %. The furan fraction is introduced preferentially by furaldehydes (furfural content about ¾). Influence of a preliminary heating is most significant at the radiolysis initial stage - more prompt migration and volatilizing of primary products of a fragmentation takes place, as well as decomposition of thermally astable fragments. It prevents participation of primary products in the subsequent radiolytic transformations. Preheating experiments reveal dominating value of radiolytic processes in formation of final products of distillation. Rough chain formation of the carbonic gas and other radiolytic products massive attenuates excessive overheat of a cellulose feedstock, thereby preventing development of destruction and distillation via the adverse "pyrogenous" mechanism.

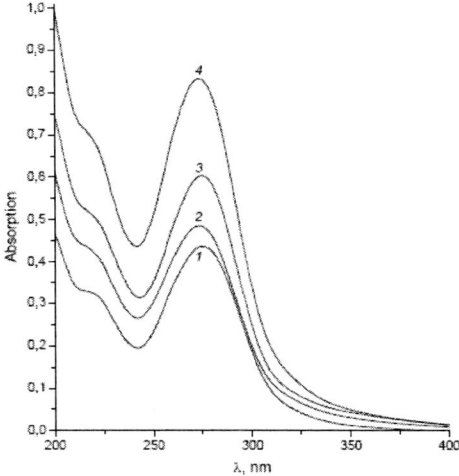

Figure 7. Optical absorption spectra of condensates originating from C2 at initial temperature 16°C (*1* - Z=49, *2* - Z=70, *3* - Z=104 cm^2/g) and 230°C (*4* - Z=104 cm^2/g).

The basic gaseous products of destruction are CO_2, H_2 and CO. Equimolar formation of carbonic gas and hydrogen is observed whereas CO yield is less by an order of magnitude. Stoichiometrically the -O-C-O- fraction (CO_2 predecessors) in $C_6H_{10}O_5$ glucopyranose unit is about 27 wt %. At the same time the experimentally observed total CO_2 + CO yield is 1/5 from dry mass of cotton feedstock. Undoubtedly that distinction between stoichiometric and experimental yields can be caused by incomplete decomposition of a feed stock, or by discrepancy between real feedstock composition and the classical formula of cellulose. For example, the microelement analysis of C3 sample testifies that a real atomic ratio C/O=1.8 whereas stoichiometric value should be C/O=1.2. Besides, real C3 contained mineral components about 2.0 wt %. Thus, the oxygen atoms deficit took place in explored C3 feedstock.

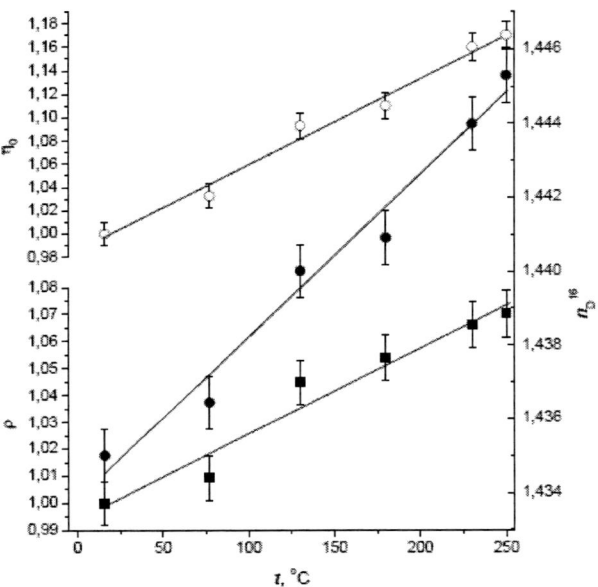

Figure 8. Influence of initial temperature of C2 samples ($Z=104$ cm^2/g) on the relative values of density ρ (*1*), viscosity η_0 (*2*) and refraction index n_D^{16} (*3*) of condensate.

Components impoverished by oxygen, possibly, were one of basic predecessors of the wood charcoal remaining in the end of distillation. At the same time, the wood charcoal part could be formed by means of condensation of destruction products into the higher-boiling compounds whose volatilization from a reaction area is complicated. The charred residue in samples C1-C3 consisted of two fractions - long fibers (structure, similar to an initial cellulose; atomic ratio C/O≈8.4; see Figure 9a) and the frothed inclusions (in the form of blobs and films; the atomic relation C/O≈6.4; see Figure 9b). The second fraction meets on surface of a charcoal and, apparently, is formed by condensation processes. In particular, condensation processes are characteristic for furfural and some its derivatives at the dry distillation conditions [2]

 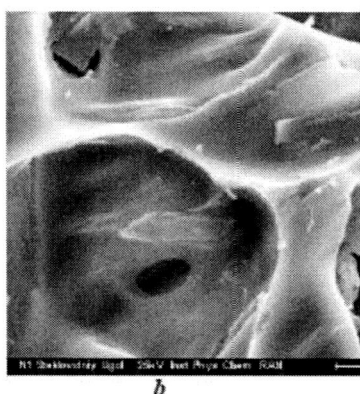

Figure 9. Two structures of charcoal (105×80 μm).

$$2C_5H_4O_2 \rightarrow C_{10}H_6O_3 + H_2O \tag{10}$$

It has been shown in [56], that furfural promptly decays in acidic medium, yielding both formic acid and humic substances (formic acid 50 g and humic substances 41.5 g are formed from furfural 100 g). Formation of high-molecular products by an interaction of furfural with other intermediate compounds [56] can also contribute to additional carbonization of feedstock. Apparently the fastest removal of vapours from irradiation area is the main method to reduce a yield of high-molecular products and to minimize losses of furfural homologues.

In practice furfural can be manufactured by acid hydrolysis of pentosans which are a part of easily hydrolyzable vegetative hemicelluloses [56].

$$C_5H_{10}O_5 - 3H_2O \rightarrow C_5H_4O_2 \tag{11}$$

In contrast to hemicelluloses, both hydrolysis and pyrolysis of cellulose are characterized by low probability of furfural formation [2, 6, 56]. The present study dealt with hemicelluloses-deficient feedstock. Hence high yield of furfural is caused just by the cellulose destruction. Taking into account real composition of samples, experimentally observed yield of CO_2 specifies that the overwhelming majority of normal glucopyranose rings decomposes during electron-beam distillation.

THE MECHANISM OF CELLULOSE DESTRUCTION VIA ELECTRON-BEAM DISTILLATION

Most likely, a radiolytic destruction of cellulose at high dose rate (electron-beam distillation) is initiated by the same reactions, as at low dose rate. However a fast beam-heating changes both decomposition paths and stabilization paths of intermediates.

Absolute majority of low-molecular-weight products of distillation contain from 1 to 6 C atoms. The composition analysis of products specifies that all of them have one predecessor - a glucopyranose ring $\sim(C_6H_{10}O_5)\sim$. Besides, the fragmentation of a glucopyranose unit can include the set of reactions - dehydration, dehydrogenation, decarboxylation, single bond cleavage, isomerization, and some other.

Reactions (2)-(7) including C-H bond cleavage on glucopyranose ring are the main ways of radiolytic formation of primary organic radicals ·R. Primary R· radicals, apparently, are instable as a result of the considerable strains induced by discrepancy between electronic configurations of radical centre (sp^2-hybridization) and initial molecular unit (sp^3-hybridization) [3]. Less stable C(1) radical easily breaks already at low temperature whereas C(4) radical decay is initiated by a heating [15].

Primary radical ·R takes allyl structure via thermostimulated dehydration, for example:

(12)

(13)

This process is detected by ESR method [3, 15, 57]. In turn, -C=C-conjugation destabilizes C-O bonds. That can also result in isolation and removal of an organic fragment or disclosing of pyranose ring. For example, the C(4) radical via reaction (7) and subsequent dehydration and decarboxylation can be transformed to a short-lived fragment which in turn can be the predecessor of furfuryl alcohol and furfural:

$$\text{(glucopyranose)} \xrightarrow{-H_2O} \text{(intermediate)} \xrightarrow{-CO_2, +H} \text{(fragment)} \longrightarrow \text{(furfuryl alcohol)} \xrightarrow{-H_2} \text{(furfural)} \quad (14)$$

$$\text{(glucopyranose)} \xrightarrow{-H_2O, -2H} \text{(intermediate)} \xrightarrow{-CO_2} \text{(fragment)} \longrightarrow \text{(furfuryl alcohol)} \xrightarrow{-H_2} \text{(furfural)} \quad (15)$$

The organic fragment liberated from glucopyranose unit can have several configurations depending on following factors: a place of C-H bond cleavage and localization position of unpaired electron in ·R radical; a position of the olefinic bond arising via dehydration or dehydrogenation; excess energy and local temperature; content and configuration of hydrogen bridges system; the steric factor controlling a removal of fragmentation products from the origin place; and some other. Presence of several shapes of the liberated fragment predetermines several paths of its stabilization.

Cleavage of C-C bonds can precede dehydration and decarboxylation of radicals. It is shown, in particular, by a low yield of water and high yield of carboxylic acids and saturated -CH$_3$ or -CH$_2$ - derivatives. Hence, the wide assortment of the final products resulting from splitting of C-C bond in the liberated fragment is observed, for example

$$\text{(glucopyranose radical)} \xrightarrow{-CO} \cdot CH_2OH + H_3C-CO-CO-CH_3 \quad (16)$$

$$\text{(glucopyranose radical)} \xrightarrow{-H} H_3C-CO-CH_2OH + O=CH-C(OH)=CH_2 \quad (17)$$

Chain propagation of cellulose destruction is caused by transformation of large primary radicals ·R into shorter secondary ones being also thermally astable. The rigid fixing of fragments by ambient molecules can interfere with their removal. In this case allyl radicals can be transformed into polyene radicals, which are more stable and precede to a charcoal formation.

$$\text{(18)}$$

Being an effective scavengers of radicals and excess energy, a part of organic fragments (for example, furfural) participates in fast reactions with other fragmentation products or decomposes - it results in expansion of assortment of final products.

Cellulose conserves a hard state and is turning yellow at low dose rate (~0.15 kGy/s) insufficient for effective heating and a distilling initiation. A condensate and charcoal yields at post-radiation (dose ~500 kGy) dry distillation are about 47 and 32 wt %, respectively. Significant decrease of a condensate yield in comparison with electron-beam distillation testifies to the important role of the reactions proceeding directly at radiolysis. Post-radiation dry distillation deals exclusively with thermal degradation of stable molecular fragments of the cellulose having reduced degree of polymerization. Probably, the intermediates participating in pyrolitic destruction process differ from products of radiolytic reactions (6) and (7). Accordingly, yields of electron-beam destruction differ from post-radiation destruction yields.

Apparently, electron-beam distillation of C2 and C3 pine cellulose samples yields different effects owing to distinction of methods of chemical pine treatment applied at cellulose production. The remaining content of lignin and hemicelluloses in cellulose depends on an applied technique of digestion and bleaching of wood feedstock [2].

Thus, the main effect of electron-beam distillation of various type of cellulose is formation of a liquid organic condensate. Furfural and its derivatives dominate among liquid products. Both crushing and a preliminary heating of a feedstock promote an increase of furans fraction in condensate. The explored mode of electron-beam distillation can become a key to development of perspective methods of furans production from widespread plant materials (including a wastage).

ELECTRON-BEAM DISTILLATION OF LIGNIN AND BINARY MIXTURES

Radiolytic destruction of cellulose as a part of wood (in situ) is less productive, than destruction of the purified cellulose fiber [58-60]. Apparently, cellulose protection in the bulk of irradiated wood is caused by an excitation energy transfer from cellulose to more stable aromatic groups of lignin [61,62].

Lignin is the second widespread vegetative polymer, however it is one of the most difficult for studying. It is polyphenolic branched polymer without regular alteration of repeating units, unlike cellulose and proteins. The primary structural units of lignin are p-hydroxycinnamic alcohols (Figure 10): p-coumaryl ($R_1 = R_2 = H$), coniferyl ($R_1 = H$, $R_2 = OCH_3$), and sinapyl ($R_1 = R_2 = OCH_3$) alcohols. The molecular chain of lignin is a combination of largely 4 types of phenoxyl radicals, of which the radical with the unpaired electron localized on the phenolic oxygen is the most significant [2]. Lignin is a waste of cellulose production and has smaller application. At the same time, annual world output of lignin exceeds 50 million tons [2, 63].

Protective action of lignin in wood was revealed via studying of the quantitative and qualitative content of radiolytic radicals [62, 64]. The irradiated wood contains both cellulose origin radicals (as in the purified cellulose fiber) and phenoxy radicals originated from lignin. In ESR spectrum of the foliar wood irradiated at 60 kGy and 77 K the signal of lignin radicals dominates in spite of the fact that about 70 % of wood mass consist of carbohydrates (cellulose and hemicellulose). In an aspen wood an yield of lignin type radicals is 1.7, whereas a yield of cellulose origin radicals is 0.8. In turn, a yield of cellulose radicals in the purified aspen cellulose is 2.7.

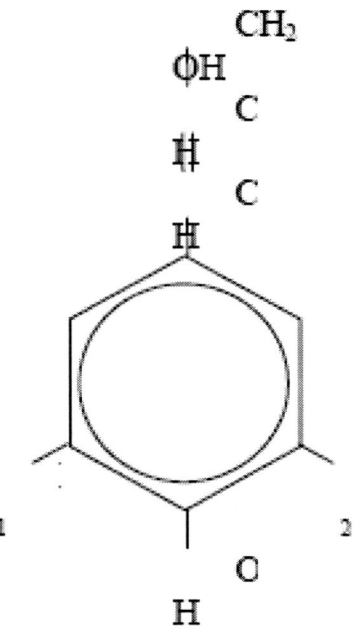

Figure 10. A primary structural unit of lignin.

Hydrolytic lignin containing 10 wt % cellulose was studied under electron-beam distillation.

The composition of the tar obtained via conventional dry distillation and electron-beam distillation of the lignin is given in the Table 7. In the case of thermal treatment, the tar consists of a wide variety of aromatic compounds with a relatively high mass fraction: the chromatogram contains about 50 peaks. The most significant components are toluene, xylenes, phenol, guaiacol, and creosol. The dry distillation products also include furfural, furanmethanols, and 5-methylfurfural. Their formation is presumably due to the cellulose admixed to the initial lignin samples.

The range of products narrows in the case of electron-beam degradation of lignin: the chromatogram of the tar contain 35 peaks, of which only 27 coincide with those of the dry distillation products [4, 5]. The base of the radiolytic tar is formed mainly by two compounds, guaiacol and creosol. However, the total yield of tar during electron-beam distillation is almost three times higher.

It is important to note that the fraction of alkylbenzols among products of radiolysis is lowered. The phenols and alkoxyphenols are the predominant liquid radiolytic products. Methane prevails among gaseous products. The formation of stable products of radiolytic fragmentation of aromatic macromolecules is determined mainly by radical processes [1, 2]. Radiolysis results primarily in the abstraction of hydrogen atoms from saturated substituents in the lignin molecule Lig.

Table 7. Main products of ordinary dry distillation (mode T) and electron-beam distillation (mode R) of lignin, wt %.

Products	Mode		
	T	R (4.8 kGy/s)	R (3.6kGy/s)
Chared residue			
Gas	56.1	48.8	46.8
Aqueous distillate fraction	20.5	15.8	18.9
Tar: total	16.2	16.8	15.3
including (wt % of tar):	7.2	18.6	19.0
Methanol + Ethanol			
Benzene	0.5	1.0	0.4
Methylbenzene (toluene)	0.8	-	-
Furfural	8.7	0.2	0.2
Furanmethanols (total)	0.5	0.1	0.2
Ethylbenzene	0.8	0.2	0.1
Dimethylbenzenes (total)	0,6	0.1	0.1
(1-Methylethyl)-benzene	9.7	0.1	0.1
Phenol	1.1	-	-
Ethyl-methylbenzenes (total)	12.4	3.6	2.9
4-Methyl-1-methoxybenzene	2.2	0.1	0.1
2-Ethyl-1,3-dimethyl-benzene	0.3	0.1	0.1
Methylphenols (total)	1.1	0.1	0.1
Methoxyphenols (total, including	2.7	1.1	0.9
guaiacol)	38.6	57.7	60.2
Dimethylphenols (total)	2.5	2.2	1.8
4-Methyl-3-methoxyphenol	0.4	3.9	2.8
Dimethoxyphenols (total)	0.5	3.9	2.3
4-Methyl-2-methoxyphenol (creosol)	5.8	19.8	21.3
4-Ethyl-2-Methoxyphenol (4-ethylguaiacol)	0.8	3.0	3.8

A substantial part of H atoms is produced via the dissociation of hydroxyl groups yielding phenoxyl radicals ArO· [1, 2]

$$\text{Lig (ArOH)} \rightarrow \text{ArO·} + \text{·H} \tag{19}$$

Similar phenoxyl radicals ArO· can be formed via the rupture of ether bonds

$$\text{Lig (Ar-O-Ar)} \rightarrow \text{ArO·} + \text{·Ar} \qquad (20)$$

The H atom easily reacting with an aromatic ring is being transformed into H-adduct [1, 2, 65]. Probably, an aromatic H-adducts have a low thermal stability. The decay of H-adduct takes place via cleavage of lateral C-C bond. The stable aromatic ArH molecule and the secondary R· radical are formed as a result

$$\text{Lig (ArR)} + \text{H·} \rightarrow \text{LigH·} \rightarrow \text{ArH} + \text{R·} \qquad (21)$$

Such an effect was reported for toluene [65]: alkylcyclohexadienyl radicals at elevated temperatures (<500°C) readily dissociate yielding benzene and the alkyl radical. Furthermore, it is known [1] that dealkylation products are formed with a higher yield than alkylation products in the radiolysis of alkylbenzenes.

The recombination of methyl and phenoxyl radicals gives a methoxyderivants

$$CH_3 + ArO· \rightarrow ArOCH_3 \qquad (22)$$

In turn, the ·CH_3 radical is being transformed into methane by eliminating hydrogen from lignin

$$CH_3 + \text{Lig} \rightarrow CH_4 + \text{Lig} \qquad (23)$$

or forms larger hydrocarbon by radical recombination

$$CH_3 + ·CH_3 \rightarrow C_2H_6 \qquad (24)$$

The tar distilled off via electron-beam heating is distinguished by high antioxidizing ability, in particular, in processes of inhibition of styrene thermopolymerization. The Figure 11 collates the formation kinetics of the styrene thermopolymer in the presence of 0.025 wt % various inhibitors.

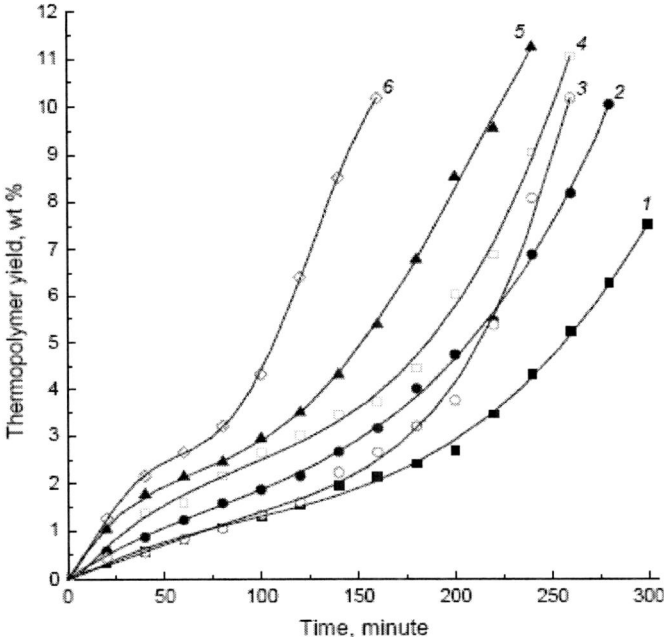

Figure 11. Polystyrene formation at 120 °C in the presence of 0.025 wt % various inhibitors: *1* - 4-dimethylaminomethyl-2,6-di-*tert*-butilfenol; *2* - tar; *3* - di(methylbenzyl)-phenols; *4* - bis-(2-oxy-5-methyl-3-*tert*-butylphenyl)-methan (Agidol-2); *5*- 2,6-di-*tert*-butyl-4-methylphenol (Agidol-1); *6* - (methylbenzyl)-phenols.

It is seen that all tested phenolic inhibitors are inferior in efficiency to the reference stabilizer Mannich base - 4-dimethylaminomethyl-2,6-di-*tert*-butilfenol (the curve *1*). Note that the structure of phenolic compounds has a considerable effect on their inhibiting activity. For example, monosubstituted phenols - methylbenzylphenols - are less effective than their disubstituted counterparts - di(methylbenzyl)phenols. The kinetic curves for the case of the di(methylbenzyl)phenol mixture (curve *3*) and for the case of the unpurified tar obtained via electron-beam distillation from lignine (curve *2*) are closer to the reference curve 1 than other curves. The inhibitory effect of commercial stabilizers Agidol-1 (2,6-di-*tert*-butyl-4-methylphenol) and Agidol-2 (bis-(2-oxy-5-methyl-3-*tert*-butylphenyl)-methan) was noticeably weaker than that of the tar. Moreover, the intermixture of tar with the Mannich base demonstrates a synergistic magnification of the inhibition ability (see Figure 12).

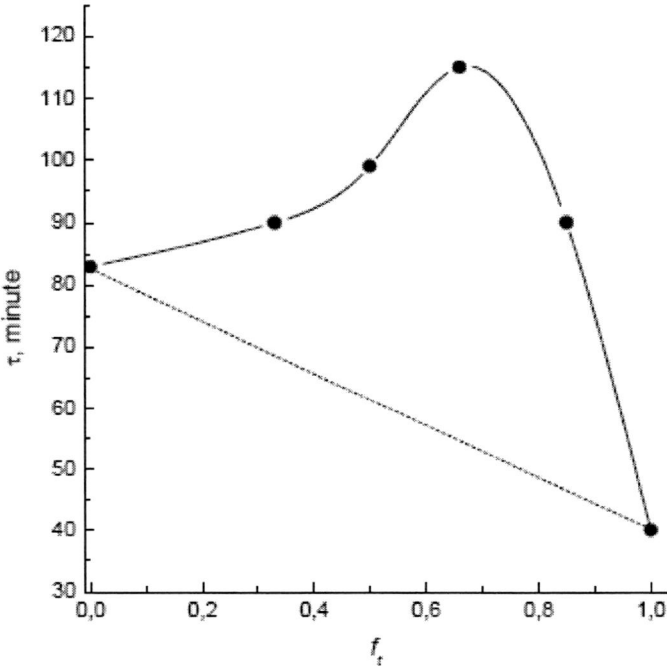

Figure 12. Induction periods τ of styrene thermopolymerization at 120°C in the presence of 0.05 wt % the mixed inhibitor consisting of tar and Mannich base (f_t - the tar fraction in the mixed inhibitor).

High inhibitory effect can testify to presence of polyhydroxy phenols in tar composition. Such components dominate in the fraction boiling at $t \geq 235$ °C. Thus, the tar generated by electron-beam distillation of wood or lignin can be applicable for practical production of polymerization retarders.

Lignin or lignocellulose affects productivity of electron-beam distillation of binary mixtures containing a synthetic polymer or a heavy oil hydrocarbon as the second component. Effective cracking of heavy hydrocarbons can be initiated by radiation [1], in particular, by electron-beam radiation [66]. At a temperature of $T \geq 300$°C, the radiolytic cleavage of bonds in hydrocarbon macromolecules is accompanied by the thermally stimulated chain destruction of generated radicals. The advantage of the radiation-initiated thermal cracking is the effective formation of radicals at temperatures below those required for conventional thermal cracking [1, 66]. Electron-beam destruction of wood components is controlled also by the chain mechanism. It is of interest to trace the simultaneous occurrence of two radiation-initiated chain processes using wood–bitumen mixtures as an example.

The electron-beam distillation of one-component samples differs in the rate — lignocellulose feedstock is distilled several times faster than the natural bitumen. In the study of bitumen and binary samples, the yields of distillation products were measured at a dose $D = 300$ kGy; i.e., within the dose sufficient to complete the distillation of lignocellulose, but insufficient to complete the distillation of individual bitumen. Irradiation was conducted at initial temperatures T_0 of 18 and 200°C. The observed dependences of the yield of liquid products of distillation on the feedstock composition are presented in the Figure 13.

At $D = 300$ kGy and $T_0 = 18°C$, about 60 and 28 wt % of organic liquid is distilled from pine wood and lignin, respectively. In turn, the yield of low-molecular condensate from the bitumen is 37 wt %. If the distillation of degradation products from binary mixtures of bitumen and wood sawdust obeyed additive rule, the total yield of distilled products G_Σ would decrease with an increase in the proportion of bitumen in the mixture (as shown by dotted line). However, the data in the Figure 10 show that G_Σ is higher than it should be expected from the additive rule. A similar effect is observed in mixtures of bitumen and lignin. At an initial temperature of the mixtures of $T_0 = 200°C$, the dependence of G_Σ on the composition also indicates the presence of a synergistic effect.

The propagation step of the chain degradation process for bitumen hydrocarbons $R\dot{C}_nH_{2n-1}R'$ can be conventionally represented as a set [1] of radical reactions of hydrogen abstraction (25) and β-cleavage (26)

$$\dot{C}_mH_{2m}R + RC_nH_{2n}R' \rightarrow RC_mH_{2m+1} + R_i\dot{C}_nH_{2n-1}R' \tag{25}$$

$$R\dot{C}_nH_{2n-1}R' \rightarrow R\dot{C}_kH_{2k} + R'C_{n-k}H_{2(n-k)-1} \tag{26}$$

in which small radicals abstract hydrogen from a heavy hydrocarbon, and large radicals dissociate to alkenes and a small radical. The termination of the chain process is caused by dimerization or disproportionation of radicals.

It is quite likely that small intermediates of wood origin can be converted into final low-molecular products via the abstraction of hydrogen from bitumen components (similar to reaction (25)), thus contributing to the development of the thermally stimulated chain decomposition of hydrocarbons.

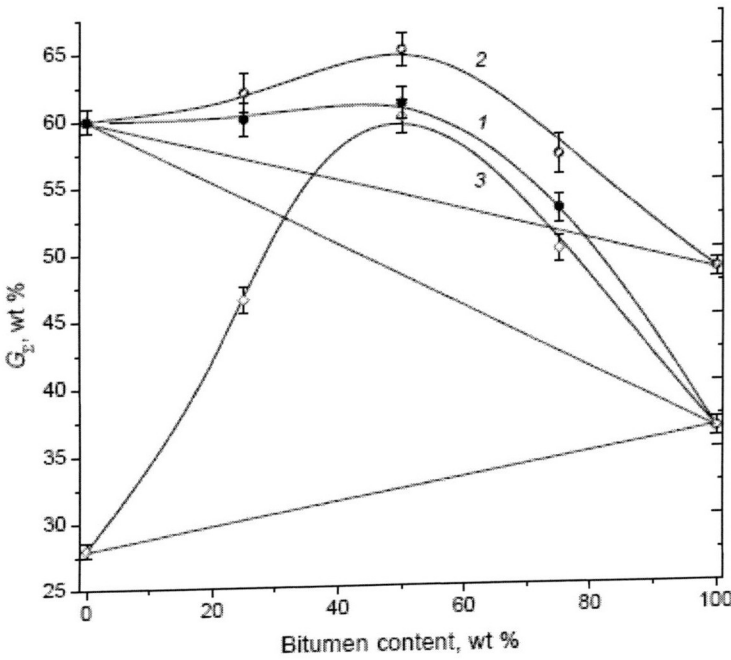

Figure 13. The dependence of the total yield of condensate G_Σ on the bitumen content in mixtures with (*1, 2*) pine sawdust and (*3*) lignin at initial feedstock temperature (*1, 3*) 18 and (*2*) 200°C.

The composition of products derived from the binary mixtures indicates the formation of compounds, the origin of which can be explained by cross combination of intermediates of wood and bitumen origin - alkyl acetates, phenyl acetates, alkylfurans, benzofurans, etc. The yield of these "hybrid" products makes up to 15% of the total yield of condensate.

Along with the participation in the chain transfer reaction, wood components can play a more sophisticated role. In particular, the starting binary mixtures have loose consistency, which could facilitate the distillation of bitumen degradation products from the irradiation area. In addition, light products of wood degradation can facilitate the distillation of heavy hydrocarbons owing to the "steam distillation" effect. Charcoal produced from the binary mixtures is crumbly, does not stick together, and is easy to withdraw from the reactor, properties that compare it favorably with conventional coke produced by the cracking of bitumen.

It is important that wood and lignin degradation products act as inhibitors of radical polymerization [10]. They impede tarring in the bitumen fraction. For example, upon the distillation of binary mixtures, in which bitumen was replaced with polystyrene, we observed an almost complete recovery of styrene, which is usually easy to polymerize by heating. Increase of radiolytic destruction yield of synthetic polymers was observed also in the wood-polymer mixtures containing polyethylene-terephthalate, polyvinyl-acetate, polyethylene, polypropylene and some other.

Thus, under the conditions of electron-beam heating and distillation of binary mixtures of natural bitumen or synthetic polymer with wood, the synergistic increase in the yield of condensable products of their degradation has been revealed. The effect is due to an increase in the efficiency of the decomposition of mainly the non-wood fraction.

CONCLUSION

Thus direct productive conversion of wood into valuable chemicals can be implemented by means of the electron-beam distillation. Technical application of this method can be considered as reasonable prospect because electron-beam distillation is characterized by valuable performances: the chain mechanism, high yield of a organic condensate, selectivity, low enough energy consumption, and uniprocessing. Electron-beam conversion of wood into organic liquid can be interesting as initial stage for subsequent formation of various products in industry: effective polymerization retarders; monomers and reagents for polymeric production; the alternative liquid fuel; special types of wood charcoal; and some others.

Polymerization retarders. Phenolic retarders of polymerization are applied for a long time in the industry, in particular at styrene processing. Phenols of a coke-chemical and petrochemical origin are used more often in the today's polymeric industry. Vegetative lignin is the major reproducible source for phenols production. Conventional dry distillation of lignin or wood yields a mixture of benzenes and phenols whereas electron-beam distillation produces exclusively phenols (including dibasic phenols). It has been shown above that the crude tar obtained by distillation of lignin has higher inhibiting effect in comparison with many other industrial inhibitors. The same tar immixed with 4-dimethylaminomethyl-2,6-di-*tert*-butilphenol yields synergetic inhibiting effect in styrene. High yield of the active aromatic tar allows to regard the electron-beam distillation of lignin and coniferous wood as one of competitive paths for production of industrial polymerization retarders, antioxidants and stabilizers of monomers.

Monomers. Many household plastics easily decompose under the influence of electron-beam processing in the presence of lignocelluloses. Monomers or the low-molecular-weight oligomers formed as a result can be distilled off at stable state. Lignocelluloses destruction products react as inhibitors interfering with backward polymerization reactions. Such effect can be applied to processing of a complex domestic wastes (paperstock and plastics) for the purpose of regeneration of low-molecular-weight chemical raw materials. At the same time, the condensates forming by electron-beam distillation of lignocelluloses consist also of the molecules having olefin bonds and the active carbonyl groups. Similar substances are widely used both as monomers and as auxiliary reagents in various polymerization processes [67]. In the present work effective formation of polymers was observed at a heating or boiling of the condensates originated from purified cellulose. As the yield of condensates from woods is more than 50 wt % the electron-beam distillation attracts attention as a productive mode for prospective formation of useful monomers and reagents.

The alternative fuel. Liquid products from electron-beam distillation cannot be used directly as fuel. Presence of nonsaturated bonds and the active functional groups promotes undesirable chemical interaction (dehydration, polyfunctional condensation, etc.) between components in condensates. Alkanes, benzenes, aliphatic alcohols, ethers and esters are usually perceived as suitable components for liquid fuel [68]. Tetrahydrofuran derivatives are also considered as a proper component for high-octane engine fuels [69].

Our experiments indicate that hydrogenation and/or alkylation of organic vapors from electro-beam distillation of wood feedstock can yield stable fuels (motor, diesel, reactive or boiler fuels). The radiolytic variant of such transformation follows from observed influence of gas flow rate on condensate composition [6]. Participation of gaseous alkanes or hydrogen in radiolytic destruction of a vegetative feedstock results in several effects. Gas flow facilitates removal of fragmentation products from feedstock being irradiated. High dilution of vapour by the gas prevents both secondary processes of vapour fragmentation and undesirable reactions between vapour components. Hydrogenous containing molecules RH used as carrier gas contribute to the radiolytic formation of ·H atoms and alkyl radicals ·R [1, 70]

$$RH^* \rightarrow \cdot R + \cdot H \qquad (4)$$

initiating both chain decomposition of feedstock and processes of hydrogenation and alkylation of the fragmentation products

$$=C=C= + \cdot R \rightarrow =CR-\dot{C}= \qquad (27$$

$$=C=O + \cdot R \to =\dot{C}-OR \tag{28}$$

As a consequence the hydrocarbonaceous part in vapour molecules elongates but the vapour unsaturation degree (instability) is being decreased. At last, radiolysis of the gas results in a condensates enrichment by valuable fuel components - branched alkanes [71]. The intermediates formed from carrier gas react preferentially with unsaturated vaporous as scavengers therefore the utilization yield of gas also increases due to prevention of its regeneration.

Earlier we have shown [6] that the birchwood in a methane flow can be transformed to the liquid engine fuel having proper fractional composition and octane value > 90. High dose rate (to prevent synthesis of high-molecular compounds) and low concentration of vapours in a vapour-gas mixture (to preclude polycondensation) are the important parameters needed to formation of qualitative fuel.

As it has been shown above, electron-beam processing of native bitumen or high-viscosity oil in the presence of lignocellulose can be used too in fuel production. Destruction of lignocelluloses influences comprehensively the destruction of heavy hydrocarbons, incrementing a yield of easy fragments at rather small heating.

Wood charcoal. Conventional dry distillation yields the wood charcoal essentially differing on composition and structure from the chared residue formed by electron-beam distillation [5]. Carbonification is incomplete in both modes. But electron-beam mode produces wood charcoal, conserving an ordered structure of initial cellulose fibers and lignin aggregates. In turn, conventional dry distillation results in charcoal deformations and a sintering. The structural distinctions of wood charcoals can appear interesting from the point of view of development of new sorbents and catalytic agents.

Thus, the present work demonstrates possibility of effective transformation of lignocellulose matters into liquid organic products by electron-beam distillation. Conventional dry distillation results in large amount of both water and charcoal whereas the electron-beam heating produces much more organic liquid. Fermentative formation of bioethanol and biosolar oil is also less productive than electron-beam conversion of a vegetative feedstock into organic liquid. Besides, both enzymatic and hydrolytic processes result in large volume of wastewater, its handling is extremely labour-consuming and expensive [1]. A residual product of the discussed electron-beam conversion is wood charcoal - ordinary and convenient hard fuel or a sorbent.

High efficiency of electron-beam distillation can be a key to development of wasteless methods for plant materials transformation into the useful products, such as polymerization retarders, monomers or the alternative fuel.

ACKNOWLEDGMENT

This work was supported within the Program "Fundamental Problems in Power Industry" (principals - academician Moiseev I.I. and academician Myasoedov B.F.) of the Presidium of the Russian Academy of Sciences. Authors express gratitude to academician Tsivadze A.Yu. for the help and support.

REFERENCES

[1] Woods, R. J. and Pikaev, A. K. *Applied Radiation Chemistry: Radiation Processing; Willey Interscience*: New York, 1994.
[2] Fengel, D.; and Wegener, G. Wood *(Chemistry, Ultrastructure, Reactions); Walter de Gruyter*: New York. 1984.
[3] Ershov, B. G. Russ. *Chem. Rev.* 1998, 67, 315.
[4] Chulkov, V. N.; Bludenko, A. V.; Ponomarev, A. V. *High Energy Chem.* 2007, 41, 470.
[5] Ponomarev, A. V.; Bludenko, A. V.; Chulkov, V. N.; Liakumovich, A. G.; Yakushev I. A.; Yarullin, R. S. *Mendeleev Commun.* 2008, 18, 156.
[6] Ponomarev, A. V.; Tsivadze, A. Yu. *Doklady Phys.* Chem. 2009, 424, 57.
[7] Ponomarev A. V. *Radiat. Phys. Chem.* 2009, 78, 345.
[8] Obryadchikov, S. N. *Production of Motor Fuels; Gostoptechizdat*: Moskow. 1949 (in Russian).
[9] Byakov, V. M.; Nichiporov, F. G. *Intratrack chemical processes; Energoatomizdat*: Moscow. 1985 (in Russian).
[10] Shalyminova, D.P.; Cherezova, E.N.; Ponomarev, A.V.; Tananaev, I.G. *High Energy Chemistry*. 2008, 42, 342.
[11] Ponomarev, A. V.; Makarov, I. E.; Ershov, B. G.; Tsivadze, A. Yu. Doklady *Phys.* Chem. 2007, 416 (part 2), 271.
[12] Ponomarev, A.V.; Makarov, I. E.; Saifullin, N. R.; Syrtlanov, A. Sh.; Pikaev, A. K. *Radiat. Phys.* Chem. 2002, 65, 71.
[13] Bludenko, A.V.; Ponomarev, A. V.; Chulkov V. N. *High Energy Chem.* 2009, 43, 83.
[14] Kusama, Y.; Kageyama, E.; Shimada, M.; Nakamura Y. *J. Appl. Polym. Sci.* 1976, 20, 1679.
[15] Kuzina S.I., Mikhailov A.I. Russ. J. *Phys. Chem.* A. 2006, 80, 1874.

[16] R.Imamura, T.Uena, K.Murukami. *Bull. Inst. Res.* Kyoto Univ. 1972, 50, 51.
[17] Sultanov, K.; Azizov, U. A.; Usmanov H. U. Doklady *Phys. Chem.* 1989, 309, 907.
[18] Sakurada, I.; Okada, T.; Kaji, K. *J. Polym. Sci., Part C, Polym.* Lett. 1972, 37, 1.
[19] Ershov, B. G.; Samuilova, S. D.; Petropavlovskii, G. A.; Vasileva, G. G. Doklady *Phys. Chem.* 1984, 274, 102.
[20] Ershov, B. G.; Isakova, O. V.; Matyushkina, E. P.; Samuilova, S. D. *Cellul. Chem. Technol.* 1987, 21, 331.
[21] Ershov, B. G.; Komarov, V. B.; Samuilova, S. D. *Cellul. Chem. Technol.* 1987, 21, 457-.
[22] Klimentov, A. S.; Petropavlovskii, G. A.; Kotelnikov, N. E.; Skvortsov, S. V.; Volkova, L. A. *Cellul. Chem. Technol.* 1985, 19, 251.
[23] Ershov, B. G.; Isakova, O. V.; Matyushkina, E. P.; Samuilova, S. D. *High Energy Chem.* 1986, 20, 142.
[24] Blouin, F. A.; Arthur, J. C. *Textile Res. J.* 1958, 29, 198.
[25] Arthur, J.; Blouin, F. A.; Demint, R. *J. Am. Dyes Rep.* 1960, 49, 383.
[26] Burczak, K.; Pekala, W.; Rosiak, J. Wiad. *Chem.* 1983, 37, 193.
[27] Bludovsky, B.; Duchacek, V. J. *Radiochem. Radioanal.* Lett. 1979, 38, 21.
[28] Duchacek, V.; Bludovsky, R. J. *Radiochem. Radioanal.* Lett., 1979, 38, 31.
[29] Bludovsky, R.; Prochazka, M.; Kopoldova, J. J. *Radioanal. Nucl. Chem.* 1984, 87, 69.
[30] Komarov, V. B.; Samuilova, S. D.; Kirsanova, L. S.; Morozov, V. A.; Kuleshova, T. M.; Smirnov, A. G.; Ershov, B. G. Russ. *J. of Appl. Chem.* 1993, 66, 393.
[31] Horio, M.; Imamura, R.; Murukami, K. Butt. *Inst. Res.* Kyoto Univ. 1965, 43, 117.
[32] Dilli, S.; Garnet, J. L. *Chem. Ind.* (London). 1963, 409.
[33] Nekhaichuk, A. D.; Evdokimov, A. M.; Kitaev, S. Kh.; Moskaleva, V. E.; Yatsenko-Khmelevskii, A. *A. Chem. of Wood* (Russian). 1974, (1), 32.
[34] Freidin, A. S. *Action of ionizing radiation on wood and its components;* Nauka: Moscow, 1961; 75.
[35] Florin, R.; Wall, L.; Brown, D. *Trans. Faraday Soc.* 1960, 56, 1304.
[36] Makatun, V. N.; Potapovich, A. K.; Ermolenko, N. N. *Polymer Science.* 1963, 5A, 467.

References

[37] Baugh, P. J.; Hinojosa, O.; Arthur, J. *J. Appl. Polym. Sci.* 1967, 11, 1139.
[38] Arthur, J.; Hinojosa, O.; Tripp, V. W. *J. Appl. Polym. Sci.* 1969, 13, 1497.
[39] Arthur, J. C.; Mares, T.; Hinojosa, O. *Textile Res. J.* 1966, 36, 630.
[40] Dilly, S.; Ernst, J. T.; Garnet, J. Aust. *J. Chem.* 1967, 20, 911.
[41] Worlungton, K.; Baugh, P. *Cellul. Chem. Technol.* 1971, 5, 23.
[42] Hinojosa, O.; Nakaraura, Y.; Arthur, J. *J. Polym. Sci.*, Pan C. Polym. Lett. 1972, 37, 27.
[43] Siraionescu, Cr.; Butnaru, R.; Rosmarin, Gh. *Cellul. Chem. Technol.* 1973, 7, 153.
[44] Khamidov, D. S.; Azizov, U. A.; Milinchuk, V. K.; Usmanov, H. U. *Polymer Science.* 1972, 14A, 838.
[45] Shimada, M.; Nakamura, Y.; Kusama, T.; Matsuda, O.; Kageyama, E. *J. Appl. Polym. Sci.* 1974, 18, 3387.
[46] Gaponova, I. S.; Pariiskii, G. P.; Toptygin, D. Ya. *Polymer Science.* 1977, 19B, 706.
[47] Ershov, B. G.; Klimentov, A. S. *Polymer Science.* 1977, 19A, 808.
[48] Klimentov, A. S.; Ershov, B. G.; Bykov, A. E. *Chem. of Wood* (Russian). 1977, (2), 74.
[49] Plotnikov, O. V.; Mikhailov, A. I.; Payavee, E. L. *Polymer Science.* 1977, 19A, 25.
[50] Hon, N.-S. *J. Macromol. Sci.* Chem., 1976, A10, 1175.
[51] Golova, O. P. *Russ. Chem.* Rev. 1975, 44, 687.
[52] Takeshi, S.; Shosaki, K. *Nucl.* Eng. (Jpn.), 1977, 23, 41.
[53] Ershov, B. G.; Komarov, V. B. *Polymer Science.* 1985, 27B, 132.
[54] Ershov, B. G.; Komarov, V. B.; Samuilova, S. D. *Polymer Science.* 1985, 27B, 430.
[55] Ponomarev, A. V.; Bludenko, A. V.; Chulkov, V. N.; Tananaev, I. G.; Myasoedov, B. F.; Tsivadze, A. Yu. *High energy Chem.* 2009, 43, 350.
[56] Scherbakov, A. A. Furfural; Kiev: Gos. Izd-vo Tekhn. *Lit-ry USSR*, 1962 (in Russian).
[57] Kuzina, S. I.; Mikhailov, A. I. *Russ. J. of Phys. Chem.* A. 2005, 79, 1115.
[58] Arnapalskii, I. N.; Skrigan, A. I.; Kuprina, N. S. *Chem. of Wood* (Russian), 1975, (1), 126.
[59] Sergeeva, V. N.; Kreitsberg, Z. N.; Yakobsone, M. Ya. *Chem. of Wood* (Russian). 1976, (5), 58.
[60] Ficher, K.; Goldberg, W.; Wilke, M. *Lenzinger Ber.* 1985, 33.

[61] Campbell, F.J. *Radial. Phys.* Chem. 1981, 18, 109.
[62] Ershov, B. G.; Isakova, O. V.; Komarov, V. B.; Matyushkina, E. P. *Chem. of Wood* (Russian). 1986, (1), 6.
[63] Zakis, G.F. *Functional Analysis of Lignins and Their Derivatives.* Zinatne: Riga, 1987 (in Russian).
[64] Kuzina, S. I.; Demidov, S. V.; Brezgunov, A. Yu.; Poluektov, O. P.; Grinberg, O. Ya.; Dubinskii, A. A.; Mikhailov, A. I.; Lebedev, Ya. S. *Physics and chemistry of the elemental chemical processes.* Reports of 5th All-Russia Conference. Chernogolovka. 1997, 310 (in Russian).
[65] Amano, A.; Horie, O.; Hanh, N.H. *Int. J. Chem. Kinet.* 1975, 8, 321-340.
[66] Bludenko, A.V.; Ponomarev, A.V.; Chulkov, V.N.; Yakushev, I.A.; Yarullin, R.S. *Mendeleev Commun.*, 2007, 17, 227.
[67] Sykes, P. *A guidebook to mechanism in organic chemistry*, Longman: London. 1971.
[68] Ponomarev, A.V. *Mendeleev Commun.* 2006, 5, 256.
[69] Paul, S. F. *Alternative Fuel*, US Pat. No. 5,697,987; 1997.
[70] Cserep, G.; Gyorgy, I.; Roder, M.; and Wojnarovits; L. *Radiation Chemistry of Hydrocarbons*, Akademiai Kiado: Budapest. 1981.
[71] Ponomnarev, A. V.; Tsivadze, A. Yu. *Dokl. Phys.* Chem. 2006, 411 (part 2), 345.

INDEX

A

absorption spectra, 33
abstraction, 43, 47
accelerator, 10
acetic acid, 11
acetone, 11
acid, 19, 24, 30, 35
acidic, 35
additives, 20
aging process, 21
alcohols, 20, 41, 52
aldehydes, 20
alkane, 20
alkenes, 47
alkylation, 44, 52
alternative energy, 10
antioxidant, 17
aromatic compounds, 21, 42
aromatics, 19, 21
atmospheric pressure, 14, 18, 19, 20
atoms, 20, 25, 33, 36, 43, 52
Avogadro number, 23

B

base, 7, 42, 45, 46
beams, 10
benzene, 43, 44
biofuel, 9
biomass, 9, 11
bleaching, 38
bonds, 15, 25, 36, 37, 44, 46, 52

C

capillary, 13
carbohydrates, 41
carbon, 11, 17, 25, 26
carbon atoms, 17
carbon dioxide, 26
carbonization, 35
carbonyl groups, 24, 26, 52
carboxyl, 23, 24, 27
carboxylic acid, 37
carboxylic acids, 37
cellulose, 7, 9, 11, 18, 19, 21, 23, 24,
   25, 26, 27, 28, 29, 30, 31, 32, 33,
   34, 35, 36, 38, 39, 41, 42, 52, 53
chain transfer, 48
chemical, 9, 10, 11, 23, 24, 26, 30,
   38, 51, 52, 57, 60
chemical industry, 9
chemical interaction, 52
chemicals, 9, 51
cleavage, 14, 15, 25, 26, 28, 36, 37,
   44, 46, 47
cleavages, 27, 28
CO2, 26, 35
coke, 11, 48, 51
commercial, 45
composition, 13, 17, 18, 19, 20, 31,
   33, 35, 36, 42, 46, 47, 48, 52, 53

compounds, 11, 17, 20, 24, 26, 31, 34, 35, 42, 48
condensation, 13, 17, 34, 52
configuration, 37
conjugation, 36
consumption, 9, 10
COOH, 24
cooling, 14, 26
copolymer, 13
correlations, 31
cotton, 24, 29, 33
crops, 9
crystalline, 23
cultivation, 9

D

decay, 15, 25, 26, 36, 44
decomposition, 18, 31, 32, 33, 36, 47, 49, 52
deficit, 9, 33
degradation, 17, 18, 20, 42, 47, 48, 49
degradation process, 47
dehydration, 36, 37, 52
deposition, 9
depth, 9
derivatives, 18, 31, 32, 34, 37, 39, 52
destruction, 7, 10, 11, 13, 14, 23, 24, 25, 26, 27, 28, 31, 32, 33, 34, 35, 36, 38, 41, 46, 49, 51, 52, 53
diffusion, 26
digestion, 38
dimerization, 47
dispersion, 29
dissociation, 43
distillation, 7, 11, 13, 14, 15, 17, 18, 20, 29, 30, 31, 32, 34, 35, 36, 38, 39, 42, 43, 45, 46, 47, 48, 49, 51, 52, 53, 54
distribution, 10, 20
dressing material, 10

E

electron, 7, 10, 11, 13, 14, 15, 17, 18, 20, 23, 25, 29, 30, 31, 35, 36, 37, 38, 39, 41, 42, 43, 44, 45, 46, 47, 49, 51, 52, 53, 54
electrons, 10, 25
endothermic, 14
energy, 7, 9, 13, 14, 15, 20, 26, 37, 38, 41, 51, 59
energy consumption, 51
energy transfer, 20, 41
environment, 9
equipment, 10, 13
ESR, 36, 41
ethers, 10, 20, 52
evaporation, 14
excitation, 15, 41

F

feedstock, 11, 14, 15, 18, 19, 20, 29, 31, 32, 33, 35, 38, 39, 47, 48, 52, 53
fertility, 9
fiber, 41
fibers, 34, 53
films, 34
fixation, 20
formation, 10, 14, 17, 20, 23, 24, 26, 27, 28, 29, 31, 32, 33, 35, 36, 38, 39, 42, 43, 44, 45, 46, 48, 51, 52, 53
formula, 33
fractional composition, 53
fragments, 21, 26, 31, 32, 38, 53
free radicals, 25
furan, 11, 18, 31, 32

G

glucose, 24

H

hardwoods, 18
harvesting, 9
heavy oil, 46
helium, 13
hemicellulose, 41
high-molecular compounds, 53
hybrid, 31, 48
hybridization, 25, 36
hydrocarbons, 9, 46, 47, 48, 53
hydrogen, 13, 25, 26, 33, 37, 43, 44, 47, 52
hydrogen abstraction, 47
hydrogen atoms, 43
hydrogenation, 52
hydrolysis, 10, 35
hydroxyl, 43
hydroxyl groups, 43

I

industry, 10, 51
inhibition, 44, 45
inhibitor, 46
initiation, 14, 38
ionization, 15, 25
ionizing radiation, 10, 58
irradiation, 7, 13, 14, 24, 27, 28, 29, 31, 35, 48
isobutane, 20
isolation, 36
isomerization, 36
isomers, 20

K

ketones, 20
kinetic curves, 45
kinetics, 44

L

liberation, 15
light, 11, 13, 20, 21, 25, 30, 48

lignin, 17, 38, 41, 42, 43, 44, 46, 47, 48, 49, 51, 53
linear dependence, 24, 28
liquids, 11
localization, 37

M

macromolecular chains, 15
macromolecules, 10, 17, 43, 46
magnitude, 20, 33
majority, 14, 35, 36
manufacturing, 10
mass, 10, 17, 19, 20, 31, 32, 33, 41, 42
materials, 9, 39, 54
matter, 4
Mendeleev, 57, 60
metallurgy, 11
methanol, 11
methylbenzenes, 43
migration, 32
modelling, 10
moisture, 23
mole, 14
molecular mass, 18, 19
molecular weight, 11, 29
molecules, 15, 20, 38, 52, 53
monomers, 7, 51, 52, 54
Moscow, 57, 58

N

nanometers, 15
nitrates, 10
nitrogen, 25

O

octane, 52, 53
oil, 7, 9, 53
oligomers, 51
optical density, 31
organic compounds, 29
ox, 20
oxygen, 20, 33, 34, 41

P

permeability, 23
petroleum, 7
phenol, 42, 45
phenolic compounds, 17, 45
phenoxyl radicals, 41, 43, 44
physicochemical properties, 10
plastics, 51
polycondensation, 53
polymer, 7, 26, 41, 46, 49
polymeric chains, 26
polymerization, 23, 26, 27, 38, 46, 51, 52, 54
polymerization process, 52
polymerization processes, 52
polymers, 11, 52
polypropylene, 49
polystyrene, 49
preparation, 4
preservation, 14
prevention, 53
primary products, 31, 32
probability, 14, 20, 31, 35
propagation, 38, 47
protection, 41
proteins, 41
pulp, 29
pyrolysis, 14, 15, 31, 35

Q

quartz, 13, 14

R

radiation, 10, 11, 13, 14, 17, 26, 28, 30, 31, 38, 46
Radiation, 13, 18, 23, 57, 60
radical pairs, 15
radical polymerization, 7, 49
radical reactions, 47
radicals, 15, 25, 26, 36, 37, 38, 41, 44, 46, 47, 52
raw materials, 7, 9, 10, 11, 52
reactions, 26, 31, 36, 38, 52
reactivity, 10
reagents, 7, 10, 11, 51, 52
recombination, 44
recommendations, 4
recovery, 7, 49
refraction index, 11, 31, 32, 34
regeneration, 7, 52, 53
resources, 9, 10
rings, 35
room temperature, 23, 24, 25
Russia, 60

S

safety, 9
sawdust, 47, 48
scavengers, 23, 38, 53
schema, 13
second generation, 9
selectivity, 51
self-organization, 14
sintering, 53
softwoods, 17, 18
solubility, 10
solution, 11
specific surface, 15, 18, 19, 20, 29, 31, 32
stabilization, 36, 37
stabilizers, 45, 51
state, 38, 51
states, 15, 25
storage, 17, 18
structure, 9, 10, 23, 34, 36, 45, 53
styrene, 44, 46, 49, 51
sustainable development, 9
synergistic effect, 47
synthesis, 7, 11, 53
synthetic polymers, 49

T

tar, 11, 42, 43, 44, 45, 46, 51
technologies, 10

temperature, 11, 14, 15, 24, 26, 28, 29, 31, 32, 33, 34, 36, 37, 46, 47, 48
thermal degradation, 10, 38
thermal destruction, 27
thermal stability, 44
thermal treatment, 42
toluene, 42, 43, 44
trajectory, 15
transformation, 11, 13, 38, 52, 53, 54
transformations, 10, 20, 24, 25, 32
transpiration, 14
treatment, 17, 28, 38

U

uniform, 9
USSR, 59

V

vacuum, 23, 25
valve, 14

vapor, 17
viscose, 10
viscosity, 11, 30, 31, 32, 34, 53
volatilization, 34

W

waste, 7, 41
wastewater, 53
water, 11, 13, 14, 17, 24, 37, 53
withdrawal, 17
wood, 7, 9, 10, 11, 13, 14, 15, 17, 18, 19, 20, 28, 34, 38, 41, 46, 47, 48, 49, 51, 52, 53, 58
wood species, 17
wool, 29

Y

yield, 7, 11, 13, 18, 19, 20, 21, 23, 26, 27, 28, 29, 30, 31, 33, 35, 37, 38, 41, 42, 44, 47, 48, 49, 51, 52, 53